AF824107

THROUGH THE DEATH-DARK VALLEY

A.J. BARLANTI

Copyright © 2021 by A.J. Barlanti.

All rights reserved. No part of this publication may be reproduced, distributed, or transmitted in any form or by any means, including photocopying, recording, or other electronic or mechanical methods, without the prior written permission of the author, except in the case of brief quotations embodied in critical reviews and certain other noncommercial uses permitted by copyright law.

Printed in the United States of America.

Library of Congress Control Number: 2021945877

ISBN	Paperback	978-1-68536-344-4
	eBook	978-1-68536-010-8

Westwood Books Publishing LLC
Atlanta Financial Center
3343 Peachtree Rd NE Ste 145-725
Atlanta, GA 30326

www.westwoodbookspublishing.com

I have traveled one trillion miles
To see a face I could not recognize.
I sense a smile upon lips
That will not speak for my ears to hear.
The fading of years through a kaleidoscope of stars
With pinpoints of sharp, bright colors.
I feel the heat from the suns of a western galaxy
As the glacier melts into the dance of sparkling crystals.
light years will pass by in seconds,
And the vacuum of space has no mercy—
For me there is nothing but starlight.
I am alone . . . outer space is cold . . . final;
And there is no sound—except for my heartbeat,
And there is no alone—to match that in space.

While the stargazers watch,
From so far away
To see a glimpse of light,
Of my so very distant star.

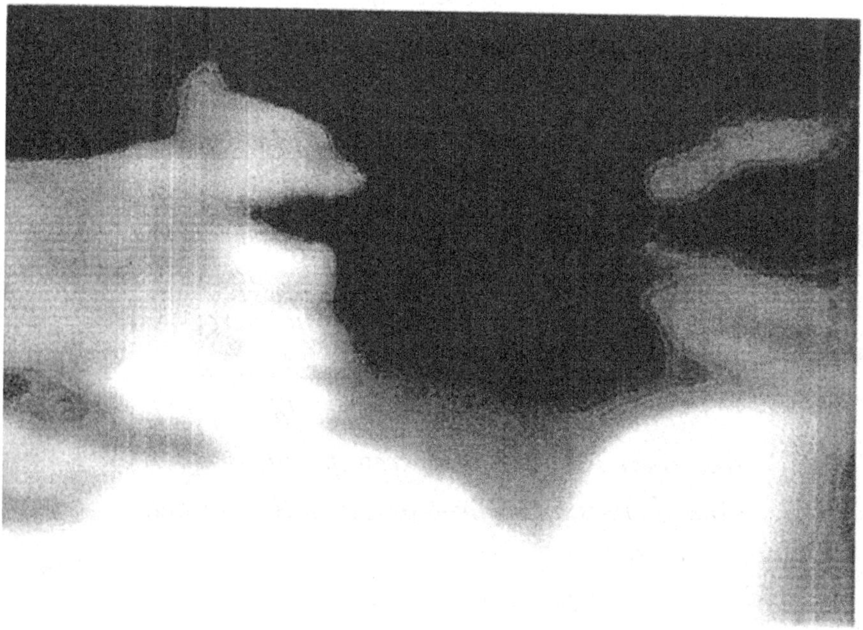

SECTION 1
PLANET "IWO TO"

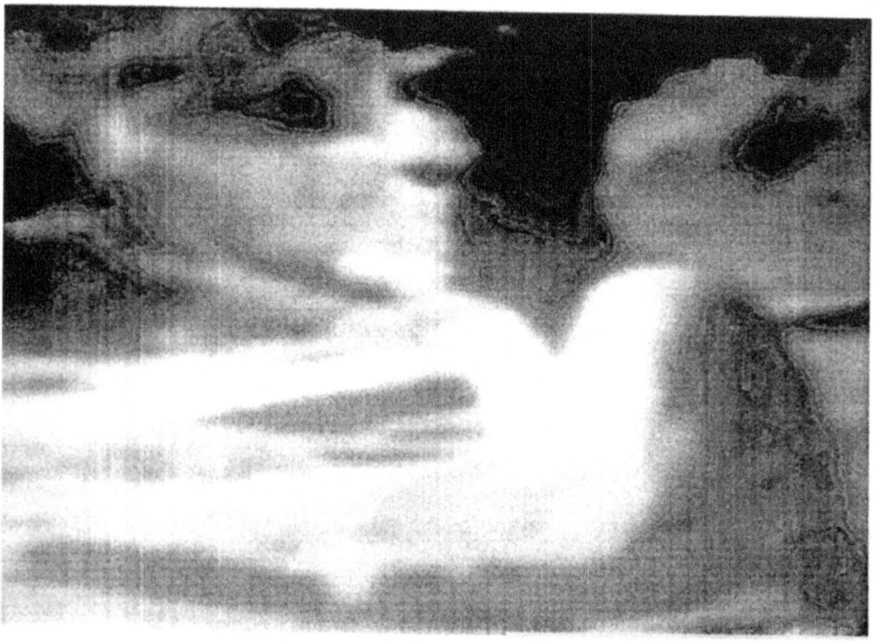

A leap of imagination to an alternate reality

THEY LANDED BY CRAFT sometime in February. It was such a barren place; a lifeless, motionless landscape of craters and rock. At first glance, a wasteland where it seemed no life could exist. Jagged stones and caves banked against waterless ravines. There were canyons of foreboding cliffs looming from above. The impression was ever-present, one that would lend no doubt that we were not alone here.

They exited their vessels onto the surface. The dark, volcanic ash held the men's ankles firmly. After a moment, they trudged forward as not to delay. Every nerve ending told them to return and not proceed along. If they paused, their boots would slowly sink into this gravel quagmire. This gave pain to the legs as they pushed ahead. Sometimes they would lose their footing and slip back slightly. It was difficult to keep ones balance in the quicksand of pebble and sand.

The air was thick with doom. The heaviness of breath seemed to stay in the lungs of those few to survive for months. There was eerie silence that spoke without words; the sound of the grave. These were men of honor; they could not let unknown terror turn them away. They pushed onward to their destiny. Each one to a different fate, but all to a common destiny.

AIRCRAFT SURVEILLANCE PHOTO

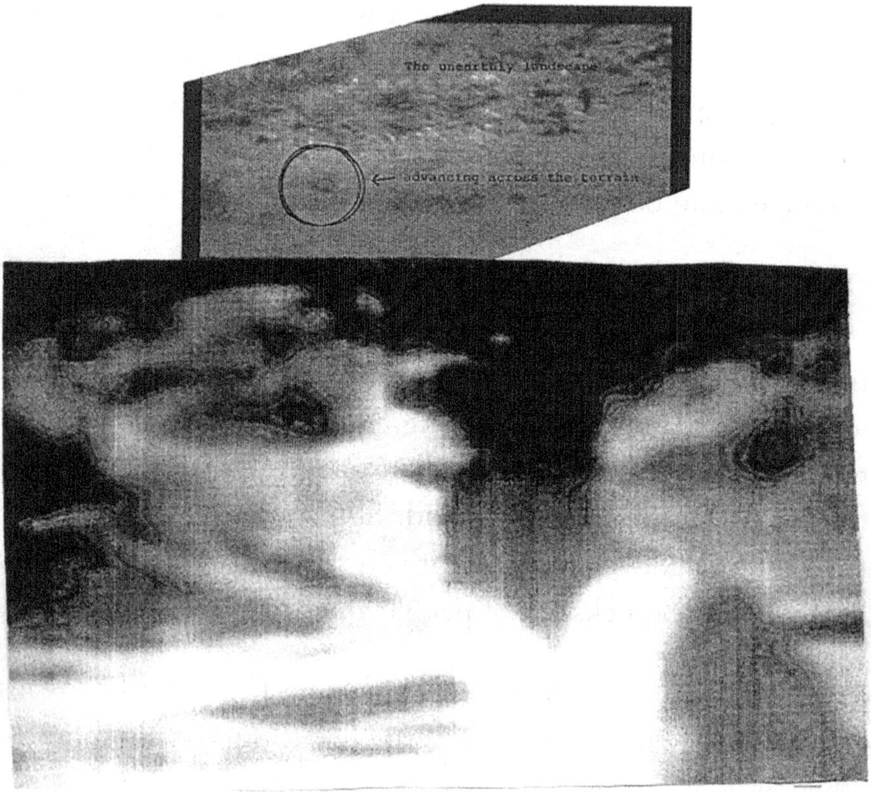

One could not shake the anticipation of an impending attack. They braced for the worst and could only hope for some hint of location from where this unknown enemy would strike. The minutes began to become oppressive. Then is that tell-tale pause before all hell lets loose.

Whistling sounded from above their heads; a shrill, screaming noise of increasing volume as it descended to the ground. Metal rained out of the atmosphere. When it crashed to the ground, ear-piercing explosions began to transpire. Bodies of the men were now being torn to shreds as the quiet was transformed into the clamor of devastation. Men yelled out in agony. The bombardment had begun.

One man carried and helped others to safety, personal risk and his own well-being did not matter. He had witnessed enough bloodshed on this day and was determined to attack those whom launched the assault. He removed his boots for better traction on this foreign terrain and continued on barefoot dodging the falling bombs. His feet would barely touch the ground for now, he would move like the wind on this soulless asteroid of land. He hurried toward the place of hidden danger ahead, while even those of exceeding valor would hesitate to advance.

This was such a strange habitat, he thought, as he saw the places from where the combatants were firing their weapons. The unseen attackers were shooting projectiles from hidden enclosures within the surface of the rocky ground. The cave-dwelling creatures were within the crevices of boulders. They were attacking from slits in the landscape to camouflage themselves.

Bombs that had fallen from the upper atmosphere exploded upon impact with the ground. The men could not defend themselves against them since they fell in random places. The metal fragments emitted from the falling devices claimed many lives. Pieces of hot metallic flame latched onto his shirt so he thought it would be better to remove it. The sulphur made the surface rock so unearthly boiling that it was fortunate it was not raining sulphuric acid as it does on the planet Venus. The exploding bombshells were causing just as many casualties and death. Hell could not inflict so much misery and human suffering as this place was. He proceeded onward in a fearless way.

The battle was underway. His helmet shifted back and forth as he went along so he thought it best to throw it to the ground to prevent the protective equipment shielding his head from blocking his view. It was an acceptable risk of head injury sacrificed for a better view of the landscape. Naturally, in this environment, a slight error in perception could cost him his life—and the lives of the men he took onto himself to defend.

He bolted though the hell-fire without becoming mortally wounded. After he moved past the point of concentrated devastation, he detected where the elusive fighters were hiding by noticing a flash from within the cliff-like structure along the terrain. He now stood poised and ready to meet the enemy head on; determined to humble the lurking foe.

The human senses amplify to cause an intense observation in an effort for self-preservation. Each moment felt as if it was the culmination of a lifetime. The five senses sharpen beyond the scope of normality—almost to the point of extra sensory. The brain is in a self-protecting mode and it shows the summit of its inactive powers. Any miscalculation could cost one's life, and at times as in this case, many other lives. When combat episodes occur, every moment is examined and savored as if it could be the last. Time slows down since the mind is racing with calculations to surroundings and thoughts of every type. Each second is experienced to the utmost with ever-increasing perception. The mind becomes photographic and almost computerized in observation.

Danger was underneath the granulated surface of this world of stone. Tunnels and a connected network of fortifications were beneath the ground. They must have been some form of underground burrowing aliens with the intention to kill. Since they could not be flushed out, the task of defeating them would cost many lives. To advance toward such an adversary takes unimaginable strength of will. To take on this type of unseen fighter alone is a testament to human courage. To annihilate in this "lone wolf" fashion is truly amazing. To silence the entire compound of eight positions single-handedly, is nothing short of incredible.

He pushed forward. Each step could be his last. He had made up his mind early on; nothing was going to stop his quest. The crouching foe must have held their breath to see his approach.

When he landed onto the curious soil of this isolated crater, he had noticed a flying craft that had crashed to the ground. His knowledge of mechanics and machinery gave him the idea of detaching the weapon from the craft. The weapon mounted in this vessel could be used to obliterate the enemy combatants. He would need to muster up all his physical strength to carry such a weapon into combat. He must first dislodge it from the craft before he could modify it for land assault. Even as he first set foot on this hollow surface, while he and his men were observing this strange area, he contemplated adapting this quick-spraying weapon. Some of parts from their accompanying weapons would combine to adapt to this superior one. Even as the carnage of battle ensued, after helping others to safety, his plan was to advance on the attacking forces. He is the type of man to entice mortal danger in his direction, away from his men. While others took cover from deadly attack, he would draw concentration in his direction, endangering himself, instead of them. This was a rare quality—an expression of loyalty combined with a type of parental protection to the degree of giving one's life to save another. His eyes pierced the foreboding horizon while the enemy prepared for a force of one.

The elusive inhabitants shielded themselves within craters and fired their weapons from behind the stone walls. They were completely entrenched, waiting for us. They had cut holes from which they could have a complete field of view. When he was fired upon, then he would know where the opposition clustered, to counter the attack. That is, if he was still alive to fight back.

The ground bursted open ten yards away, from the relentless attempted kill from the distance. He noticed the area from whence it came; now was his time. He did not verbalize this plan of offense to his men prior to this time, but as the best of warriors often do, one must act when the time is right. There is no need to talk of personal deeds of heroism.

He ran towards the flashing bursts while unleashing his roaring weapon. He pointed it to the porthole where the inhabitants huddled and ducked from his sheer ferocity of firepower. The sparks flew from out of the muzzle with great velocity. He was racing forward towards an enemy, which now were scattering away from his personal onslaught. The stone broke away under the non-stop spray perforating the outer edges of the combatants molehill structure. The sound from within their hiding spot was agonizing to the eardrum. The inhabitants fell dead and dying, he pushed his weapon into the pillbox slit- hole and continued firing. They flew about in a violent death-dance, in response to the powerful blasts.

For a moment, there was silence. He had taken them on alone, and was victorious. He was alive, unscathed to merciless attack.

All at once, the silence was shattered by repeated bursts that cracked from a cliffed peak structure to his left side. Another group of the enemy had seen his heroic charge and was intent on ending his life. About two hundred yards away, the flashes of light appeared after the stone disintegrated around the rock bunker which enhoused the hostile inhabitants. The attempt had nearly dealt him a mortal wound. He wiped the granulated dirt from his face and his feet took flight towards the enclosure as the huddled beings fired upon him through peepholes within the hollow cavities of the landscape.

He stands in full view, shows himself fearlessly, standing upright for all to witness and admire by a surprised adversary almost doubting their eyesight. The assaulting opponents were in complete awe and might have questioned the sanity of his defiant approach … but it was one of control and challenge. He seemed to be challenging both the enemy and the notion of death itself.

They had never witnessed such a display of courage. There he stood—an open target to them, almost beckoning them to try to kill him. His arm tattooed with the image of a large wildcat. He carried a portable weapon of their nightmares, one they never imagined to exist, so awe-inspiring, so lethal and rapid. They could never imagine a killing machine as this to be carried by one man. He paused for a moment, for a breath of precious life; all could now behold the spectacle of valor. Then he unleashed his frightful weapon.

The projectiles from his weapon found their mark through the rectangular slat holes of the enclosure where the enemy crept. Their screaming peppering was unrelenting, yet ineffective to disable him. The attackers were now caught up in a bee's nest of ricocheting death from within their entombed confines. He stood tall and fired away. The return fire from his modified weapon was deafening to an enemy quite mesmerized by the entire event. They discontinued their attempt as he drew nearer to them, spilling led at an unbelievable rate of speed. The metal of his weapon was smoking hot and the cloth he had wrapped around it was starting to flame up.

The enemy combatants were thrown about when hit by return fire. The wounded thrashed about while the dead lay without a pulse, motionless. He gazed at them after silencing the wounded. He had recognized these beings from a former battle. They were not tall and different in bone structure, their features were strange but they looked human. There was not any time to rest, his men were under attack. He could hear explosions from the distance. Outside the stronghold, other beings began to fire at him once again. He had almost ran out of ammunition, and would need to return across the killing fields of more concentrated fire. He must head out into the open and zigzag his way through every type of weapon pointed at him, a moving target that so far, could not be stopped.

As he made his way across the open desert of graded stone and abrasive dust, they would not wound him. Explosions generated earth-quake velocity in his pathway but he managed to avoid the terminal pricetag. Chunks of jagged pieces erupted from the ground when the shooters fired at him, but on this day, he was not to be shot down. It was as if an imaginary force-field protected him from harm, however, no such protection had been invented yet to date. Technology had created no such device at this time. Nothing in the physical realm was the cause of this, and it had gone well beyond luck and chance. It was compared to dashing through a rainstorm without getting wet. That is a more fitting description of this than the cliche 'it wasn't his day'. It was a case of reality being stranger than fiction. All the enemy's marksman fire at him and he continues to be unharmed at the end of the day. What even remains more unbelievable is his brazen approach to the hostile enclosures. He attacked as if he knew of his own invincibility.

As he traveled back to the landing point, his men were taking heavy casualties. Their mangled bodies were blown apart and the wounded were severe and critical. He helped some to medical treatment while others he carried out of harm's way. He had experienced combat before, but not like this. After seeing this devastation and loss of life, he was determined to bring this violence back to the enemy.

As he returned back to his men at the crest, he was given the additional ammunition he needed. This was only his second trip back, he would return eight times before the day was over. He was wearing the blood of the wounded men on his arms and abdomen. It was their badge of honor to his tribute. As all wars, this one was filled with dirt and blood. This was a world of sudden terror foreshadowing a seemingly quiet prelude. The lunar surface masked a stalking predator which could strike from anywhere at any time. The only constant was that the enemy was out there—ready to kill, poised for ambush from crevices unknown. The tension was constant; since they were fighting something they would often see only when it was too late.

He had felt the weight of the men he helped to safety. The gravity of the planet weighed heavily on his shoulders, to push his body down into the sinking quicksand of gravel and ash. Just to walk on this accursed gravel was turmoil for humans.

His appearance was even, young and vibrant. Many of his men were early in their years, venturing to this foreign land to risk the sacrifice of their lives. This rock of ground seemed to be such an unsettling place to die. Everyone thought this, but no one would speak the words. It was written in their eyes when they arrived.

Before long, he was moving back across the open field toward the danger of violent death, driven by a loyalty that goes against the natural instinct of self-preservation. The open range was a shooting gallery of which the targets happen to be human. He might as well draw a red bulls-eye on his back.

In an instant, a burst of led fragments hit the ground in line formation, loosening the footing beneath him. He momentarily loses his balance, then throws himself forward. From the prone fighting position, he aims for the flash point of the assaulters weapon and continues to fire into the opening porthole. When the inhabitants cease their shooting, this is time to counter attack. As he races towards the adversaries, two try to escape from above the enclosure. He wields his weapon quickly to one side, cutting them down in pieces. One of the combatants from inside the trench tries to seize the opportunity to shoot while his attention is diverted. He is quick to point his weapon back towards the opening slat at almost point blank range. The attacker is riddled with exploding wounds to blackened red clots of flesh.

He was well-beyond pity for these creatures. Even though they sometimes resembled humans, they appeared to be, at times, in a type of trance. They were first and foremost dangerous—intent to kill all of the men. They must be treated like rabid animals.

Atoms and molecules must collide in space, causing different forces of nature to react. It remains a constant law of physics. The exception to the rule is nearly impossible, so was the eight times he narrowly missed metal fragments and projectiles aimed at him. The shooting machine he customized was soon out of ammunition, since it consumed with a voracious appetite. He knew the law of average would converge sooner or later, that is, he would be hit. However, his commitment to his men and his inner strength enabled him to push onward through the open field, to once again tempt fate.

He heard shots whizzing past him within a couple of feet in each direction. Pinging sounds cracked from around all sides. He swayed from left to right, then, right to left, in an effort to make his pattern as a moving target more difficult to shoot. The assailant could not lead him into a shot by doing this.

His men witnessed his counter-offensive actions and though some were pinned down and others were dead or mortally wounded, they realized he was carrying out an unconventional counter-assault on the enemy inhabitants. He wanted to carry more ammunition on this trip, leave it about fifty yards away from his advance on the enemy when he was within range of about one hundred yards. This way, he could get back to it without having to return across the open range.

To dart across the killing fields with bulky ammunition, a weapon designed for a flying craft, on terrain with gravel that made it difficult to walk, was superhuman. To do this and not be killed, was nothing less than supernatural.

For a brief moment, there was a hesitation of the chaos of mind-shattering noise. But it lasted just for an instant, before it continued to echo and resonate near and far, with varied degrees of amplification. The pressure squeezed against the body while shock waves pressed into the skin. Some percussions could nearly knock one to the ground from yards away. It was better to anticipate the eruption, to prevent oneself from flinching. Nevertheless, after some time in this type of assault, he became conditioned to control the senses from reacting to the initial surprise of the jolt.

He could barely hear the sound of his footsteps propelling him through the black, volcanic dust. At times, he would sink in the sulphuric quicksand to his ankles. He would thrust himself forward as the ground seemed to attempt to thwart his journey. The gravel tore at his bare feet but would have collected inside his boots, making this swift run unlikely, causing him to be a slower and accessible human target.

He was driven by an individual desire, a strong will that would not yield to eminent mortal threat. He was now on a private rampage, the enemy could only watch in awe as he fought on.

The adrenaline rush throughout his body gave him almost superhuman strength. The weapon he cradled was meant to be discharged from a flying craft, not to be held and fired by one man. His agility and physical conditioning mixed with keyed-up emotional amplification allowed him to function as an unremitting fighting force. Sheer willpower compelled him along as he faced certain death with gallantry unmatched.

He awaited the hostile fire from the cliffs ahead. Judging the distance at about one hundred yards, he would flush the enemy out; baiting them by displaying himself as a human target. Their hidden position would then be revealed for his counter attack. Thus, a trap was set for those whom sought to entrap him.

A sudden burst of projectiles sounded about ten feet over his head, landing about fifteen yards in back of him. He sighted the spot from its origin and now flanked a side to continue his offensive approach. He began to close in on the cliff-like structure while firing his personally-modified weapon from the hip. The tell-tale sign that he was hitting the mark was clear—pieces broke away from the splintering edges of the recessed terrain-constructed rathole. They paused in their ambush. This was his time to zero-in on the firing port and he started his ferocious weapon.

He aimed into the rectangular-shaped opening of dug out cavern seclusion. His dreadful onslaught was about to begin. The first combatant was annihilated instantly, preparing to resume his shielded offensive. He rushed forward more steadily as his weapon sped on to devastating rapid fire. His body absorbed the undisturbed frenzy of battle, the sight of splattering aliens, the rumbling of weaponry in his arms, the atrocity of sight and sound. Bombs sounded at the landing point again, as his private battle raged on.

He was now firing into the refuge at point blank. Fluid spurted from the neck area of one assailant. Another made gasping sounds from a distorted position against the inner wall of the hillside opening. While yet another murmured indistinctly in a low tone, one sprung up from a nearby ground hole to attack. It is possible, that this one was giving the entrenched group additional cover fire. Fortunately, the modified carried weapon hit with such power, that it tore the combatant's arm off as the inhabitant's weapon broke apart.

As he thrust his sizzling weapon into the aperture, another readied to hurl a metal explosive device in his direction. He instinctively turned and blasted away. He heard the ravings of a foreign language, as the enemy fell. As he threw himself against an enormous cavern formation along the cliff-side, a terrific explosion terminated all matter in the hollow cavity; the clandestine structure had been reduced to rubble. Sand and dust clouded the vision for a minute while he wiped gravel from his forehead.

When he returned to the landing base, a crawling vehicle had been transported to enable his men to defeat this combative foe, inhabitants of this strange area that remain within the confines of its surface. The vehicle fired powerful missiles from its upper hatch. One earth- trembling explosion could help to dislodge these creatures from their hiding or drive them deeper into the networks of underground tunnels. It was quite evident that they were prepared for some time for this type of fighting.

Re-enforced, some of his men would accompany him on the journey, since he had already experienced the terrain and knew some of the deceptive whereabouts of the enemy. At a certain point, although, he would flank to one side as to draw enemy fire in his direction, instead of theirs. He decided to advance forward alone, telling his men to stay back. He would bait the unseen foe again as a human target, but this time he would have the help of the iron vehicle.

As the vehicle crept forward some yards behind, the assailants made their fateful attempt. Once he was fired upon, narrowly missing him, the hideout could be visually located. The vessel then moved into firing position. He fired relentlessly, not worrying as much about exhausting his ammunition as they returned fire. A volley of shooting sounded until the vessel dealt a decisive blast. The entrenchment split apart, as the enemy fall into view, for once on same grounds with him. Four of the enemy were still in fighting mode; the rest lay dying or dead. He drew nearer.

Two of the combatants raised their weapons to aim at him. Another, engulfed in flames ran towards his oncoming charge, yelling in a foreign tongue. The fourth hunched low to the ground with a pointed weapon, ready to strike. They were as he had been during the entire attack—now in plain sight.

He swiftly tilted his weapon from left to right, in one sweeping motion, spraying down the horizon. Two of the assailants were eliminated instantly, one nearly cut completely in half, from the waist down. The rate of speed it fired was unmatched to any they had witnessed before. The annihilation distorted their bodies, detaching parts of limbs and arms.

Just then, another missile hit the enemy position. What was left of them was obliterated. The explosion almost knocked him off his feet. The devastation was complete, one of many hiding locations of these dangerous creatures reduced to dust, but so many more awaited him and his men.

He motioned for them to return to their company position at the landing base. He would stay behind and watch for enemy assailants that might attack as the men withdrew from this position. He even covered the giant metal vehicle as it lumbered away. He was the last man that remained there to protect in the heroic fashion that he was so accustomed to.

He slung his arms over a metal fixture buried deep in the volcanic ash. His men looked on and observed him as he gazed into the ever-increasing twilight. The sky stuns the eye sight with a beauty that he would not ignore. Sunset was so alluring from the eclipsed shadow of a volcano fallen dormant so long ago. Mysterious tranquility. The calling from another world whispers to him, reminding him of his mortality, beckoning his spirit to merge with the atmosphere.

The hollow ground was uncompromising. Its volcanic gravel was unforgiving. This place was unfit for human life, so he looked up towards the heavens. He gazed at the constellations; to the milky way of galaxies, as a gentle tropical breeze brushed against his face.

He had lived up until twenty-three years of sunlight.

The upper atmosphere was darkening to welcome night.
The day was gone.
All must slip away to memories.
He noticed the sun named Alpha Centauri;
　　Closest to this solar system
This island rock satellite has a name
　　On this blue planet of land and sea
It is located in the South Pacific of the planet
　　Earth,
　　The year is 1945
　　The place is called Iwo Jima.

SINCE IWO JIMA WAS AN ISLAND OF VOLCANIC ASH WITH NO SHRUB OR VEGETATION FOR COVER WHILE FIGHTING COMBATANTS, THE UNEARTHLY LANDSCAPE CAUSED U.S. SOLDIERS TO CLAIM THAT THEY FELT LIKE THEY WERE ON ANOTHER PLANET.

"IWO TO' WAS THE JAPANESE GIVEN NAME FOR IWO JIMA BEFORE THE BATTLE WHICH BEGAN FEBRUARY 19TH 1945.

SECTION 2
THE INVASION:

A moment by moment account of combat

U.S MARINES LANDED ON THE
PACIFIC ISLAND AT 9:02 A.M
FEBRUARY 19TH 1945.

TONY STEIN WAS BORN in Dayton, Ohio on September 30th 1921. Son of Austrian / Yugoslavian immigrants, he attended Kiser High School. He worked at various occupations to support his mother, such as Delco tool shop to sharpen his knowledge of machinery, giving him the skills needed to improvise a weapon, allowing him to modify the Browning 1919 Machinegun from aircraft to land combat. He enlisted in the United States Marine Corps on September 22, 1942.

He became a Paramarine in the short lived Marine Parachute Regiment. When the regiment was disbanded, he joined into AM 28th Marines of the newly formed 5th MARDIV. A veteran of the Solomon Islands campaign in The Pacific, Tony Stein's heroics included saving a high-ranking officer from sniper attack, shielding the officer and eliminating the enemy shooters at Empress Augusta Bay. He was among the first members of his unit to set up a defense position away from the beach, after landing at Iwo Jima on February 19th, 1945, in the initial assault. It was on this day, he would destroy eight Japanese bunkers in a personal attack on enemy positions, charging the hidden machineguners with his blazing, personally-improvised weapon. Later that day, he saved his platoon by covering them as they withdrew to company position, in rear guard fashion.

He was awarded Congressional Medal of Honor for his single -handed stand on Iwo Jima and The Purple Heart.

The sight of it nearly humbled these hardened veterans. The vision was different from the carnage and bloodshed they had witnessed and experienced. Some felt as if they had seen it before in a dream... it grew as they came within its grasp. They were being drawn to it magnetically, as fate compels one to it. The massive entity was looming in an unsettling manner; a living, all-knowing force of nature that welcomed in a most foreshadowing glance.

Vertically the monstrosity rose up, as the Marines grimly observed from the ocean. It dared them to come forward. Angry, vengeful, beckoning to lure with secrets unspoken, heightening as the waves brought them closer to the five hundred fifty foot incline. It requested they arrive to never depart, a call to mortal danger and the dark manifestations of a world beyond, one that these men had brushed with in the many battles that preceded. It asked them to stay for a fortnight of years; to sleep at its base, as many men would eventually do. There was a precognition to the spectacle of viewing the final horizon of life. All things have taken them to this place. . . all events have culminated to this end; they were awestruck by the sight of it, entranced by its calling of whispers against the stark sea-breeze, taken by its glare of destiny. There it was... waiting ... to loom forever in its gigantic proportions throughout time and space—Mount Suribachi.

Suribachi cast its spell. Come to sleep at the foot, it called to the Marines, let the island have you forever. Let me hypnotize you with my compelling lullaby. There was a strange sense of premonition, dark and impending, as if an ancient force had been evoked. Before the fighting was over, thousands would give their lives on this volcanic island.

They would remain while others returned home. The frost of autumn would turn to snowfall at the base of Suribachi. The festive holiday celebrations were without them; family and friends, wives and children, would not see them again. Suribachi had forewarned with unspoken whispers on the sea-breeze. The men were indifferent to the intimidating sight of its dimensions; that they were expected to climb a vertical ascending skyward, along sulphuric pebble and volcanic ash with no cover from attacking Japanese, was not too much to ask from these battle-hardened Marines. There was foreshadowing of mortality in the wake of the morning mist. These were men of honor, they could not listen to such signals of danger, they would not even if every nerve ending warned of disaster, as it did when they beheld the presence of Mount Suribachi.

Suribachi awaited. It scoffed at the naval bombardment that had been attempting to soften its penetrating stare. The Marines were arriving to the unending barrage of big naval guns to prepare the beachhead for the landing. The powder-blue sky outlined the hell-fire which would enrage the silent volcano to decide to take the many lives that it would claim. It wished to display its natural might, to demonstrate its power that would dwarf man's destructive devices.

It had seemed as it had been experienced from another lifetime, the past years of his life. The saving lives of his men in other combat of the Pacific campaign, the days that went on for an eternity of minutes, and back in the Mad River of Dayton, where he saved the life of an African-American boy from drowning when he was nineteen. For a moment he was back in time to Ohio, underwater in the tranquility of his youth, taking a dare to jump from one of the river's cliffs, a daring young man was he. The serenity was over. There was no past. This was Iwo Jima.

Back in his homeland, at nineteen years-old, he swam to save a drowning child. The underwater muffled sounds distorted the reality of the moment, gave surreal overtones to the event, he could hear the bombing of Iwo ever so faintly, a future echo of sound far away. It was his first taste of heroism, the first life he had saved, that of a child. He had accomplished this before the war. How many others could make that claim? In the Pacific combat, he rescued many more.

Suribachi remained unimpressed, unmoved to valor. It was a Siren voice inviting the sailors to forewarning catastrophe. Arrive to my world, stay for years to come, as the winter lends its cold to your bones. Most of the Marines felt this inwardly, just as instinct hints of certain dangers unknown to sight or sound. The seagoing AMTRACS hurried to the shores of the island. He had witnessed Higgins Boats blown to shreds in previous landings. LVT's motored against waves, tossing the vessels recklessly. Booming big naval guns paralyzed hearing into oblivion, the island taking on so much bombarding that it was surprising it did not break into pieces. Every point on the island seemed to absorb incoming heavy assault. It was 9 AM and at 9:02, the United States Marines set foot onto the Pacific Island of Iwo Jima.

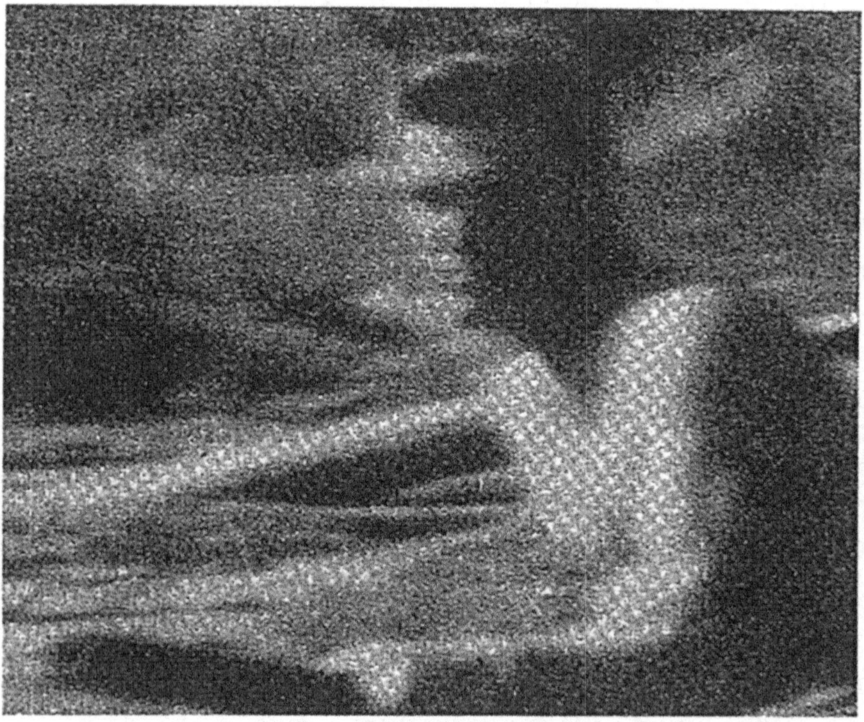

Corporal Tony Stein, while serving with Company A, First Battalion, Twenty-Eighth Marines, Fifth Marine Division is the first man on station in his unit. This was on 9:02 AM in the primary initial assault of the island February 19th, 1945.

Early on, he chose field of view by sight over personal safety, removing his combat helmet. The volcanic ash bogged the men down and made them stationary targets for the Japanese mortars and bullets. Mobility was difficult in the volcanic gravel, so he took off his army boots for better traction. He had made up his mind to storm the bunker fortifications, leading the way for his platoon and waging personal combat. The Japanese forces were hiding underground in a network of interconnecting pillboxes.

Instead of merely taking cover as the Marines were establishing ground on the beachhead of the island, Tony was contemplating his personal, lone assault. He stashed boxes of .30-caliber ammunition a few yards from the storage spot, so he could gain access easily.

He scouted ahead, since he was the first of his unit to dispatch on the sulphuric sands of Iwo. He shifted his weight from side to side as he advanced, to propel his momentum along. Tony was to become Iwo Jima's one-man demolition team; destroying a complex of pillboxes single-handedly, standing up to the hiding Japanese snipers alone, avoiding exploding mortars on all sides of him, and personally annihilating over twenty of the enemy.

The first man of his unit to be on station after hitting the beach of Iwo during the initial assault. He has armed himself with a personally land-adapted weapon of which his individual knowledge of machinery and tools enabled him to modify. He did this by unique improvisation of the Browning 1919 AN/M2 trigger. The stock and front sight from an M1 rifle was interchanged along with adding the bipod and rear sight from a B.A.R.

He was thinking of plan of attack on that morning of February 19th on an island paused as a cobra when ready to strike. When the bombing began, without hesitation, he was seen by Marines advancing through the mortars bombing around him. The exploding shells, which killed many, fell on all sides of him and his heroic dash was compared to going through a rainstorm without getting wet. When the day was over, he had accomplished this eight times, destroyed in brazed one-man style eight bunkers, directed hostile fire away from a half-track to himself, and neutralized twenty Japanese. As he returned to the beachfront each time for ammunition, he helped wounded Marines to safety. Although the aircraft Browning had twice been shot from his hands, he personally covered the withdrawal of his platoon to the company position. Standing in front of the pillboxes, one could only imagine the enemy's respect for the lone advancing adversary. To stand in plain sight, only fifty yards away, and not be struck by automatic rifle bursts, must have been quite a shock to the entrenched Japanese. If one man personifies the D-Day Invasion of Iwo, he is Corporal Tony Stein.

Tony advanced forward onto the moon-like terrain of the foreboding island of volcanic rock and blackened sand-dust. He faced this unknown territory without the company of soldiers for back-up. His private search-and-destroy mission was underway. Different rules were applicable to this chaos of enemy attack.

He surveyed the open fields for places of possible enemy bunker strongholds. One mistake in judgment could cost him his life. Tony examined the edges of the craters, the sides of the formations of rock, the darkened recesses of cliffs—for revealing signs of occupation. The pillboxes were built into the natural setting, making locating them nearly impossible to the naked eye, until the shots were fired.

Their heavy machine gun was the Type 92, mounted on bunker platforms, and the lighter Type 96, which was cumbersome, flawed but deadly. The Japanese had never even imagined a portable belt-fed machinegun coming at them, no less in the arms of just one soldier. Tony improvised the trigger personally. He had modified the weapon from an aircraft to a land assault machinegun. His invention from the ANM Browning 1919 was later named The Stinger, by the Marine Corps. He would defiantly charge the Japanese alone, spraying led at 1200-1400 rounds per minute, against all odds of survival.

Suddenly, bullets fired from a distant rock-formation against a ridged embankment. His keen observation had to be almost instantaneous, and even at that, it could often be too late. He noticed the flash point of the muzzle firing and zeroed in on the spot with his weapon. His counter attack was non-stop. Tony must close in quickly, before his ammunition was spent by the voracious appetite of the machinegun. He continued to fire into the rectangular opening of the camouflaged bunker, sending the Japanese gunners reeling, as the rounds of bullets entered like an angry hive of stinging hornets. The bullets bounced about the pillboxes as the enemy scattered. At that rate of fire, there was no time to counter the assault. The .30 caliber found its mark in the hurricane of ricocheting led. The enemy could not withstand this storm of rapid fire.

He exterminated the enemy with overpowering tenacity. Tony was unyielding. The Japanese had never suspected such resolve of purpose as from the U.S. They had not expected willpower of such from one man. The Japanese forces faced a different type of adversary, when they fought prior to the Americans, they did not estimate this type of courage and determination that they would now encounter.

The hollowed pillbox entombed the blood-drenched remains that littered the crawl-space at the back of the stone bunker. One of the enemies, leaning against the wall in a fixed position, fell to the ground while his severed head remained intact, on a make-shift wooden ledge. The dangerous assailants had to be extinguished remorselessly.

Tony had been within about ninety meters away from returning to the beach when Japanese mortars rained down. He saw his fellow Marines being torn apart as he approached where they had originally landed on the island. Bodies of men were thrown about the sharp coral rocks. They yelled out in agony for medical attention. The innards of intestines and organs glistened on the indifferent pebbled surface of a harsh island in clustered rock piles. The bomb shells created crevices after they hit, causing more difficulties to mobility, as if the volcanic heavy ash was not enough to hamper movement for an attempt at taking cover.

The wide-open asteroid appearance and void land gave no such opportunity for refuge from the merciless bombing. There was nowhere to take cover and to dig a foxhole in this stony gravel was impossible. Tony dodged the screaming shells and the flying metal fragments they discharged. Without a helmet, his reaction was spontaneous to reflex movement on the battlefield, without having to lose bearing momentarily to the tilt of his helmet. However, risk for serious mortar wound was much greater. This was a tradeoff, an acceptable risk he would undertake. His perception to instantaneous events were much more important and a slight obstruction could cost lives.

One wounded Marine raised his arm from on the ground, in an effort for recognition. Tony hoisted him to his feet, and helped him to safety across the field of exploding shells. He remembered to when he had saved the life of an officer from a sniper assault in former combat of the Pacific Islands.

They had been under attack by multiple snipers. Tony threw the high-ranking officer to the ground, shielded him from the rifle rounds, and returned fire to terminate all assaulting forces. Tony Stein had protected the military officer as The Secret Service would The President of The United States, sacrificing his body as a shield, defending with himself at greater risk, and eliminating the hostile threat. This was a testament to his value of honor and loyalty.

This attack was by far very different. The enemy was hidden everywhere and assaulting from all sides with varied weapons. The Japanese, at times would hesitate until the Marines had gotten to a certain point, before any action was taken. The enemy forces would remain silent as they would if setting a trap for an ambush. The hellfire was unrelenting and as Tony gathered his ammunition near the beachhead, after witnessing the mutilation of American bodies from the mortar shells, he was even more determined to attack them with as much relentless devastation as one man could inflict.

The 1919 AN/M2 Browning had been taken from a Douglas Dauntless Dive Bomber. The aircraft had been shot down by the Japanese. Tony would carry this weapon into combat against the fortified pillboxes. His personally modified weapon, customized for land assault, with a self- improvised trigger, was now again ready to be unleashed on the Japanese. They would not be safe in their hidden positions. Tony would find them, and destroy the entire bunker complex like a fighter plane in action.

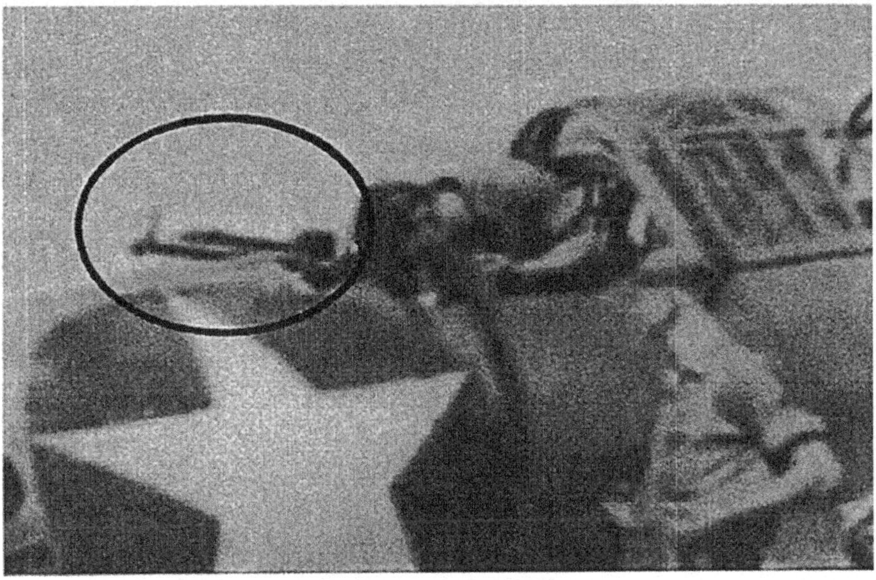

Tony Stein carried this weapon standing upright in combat at Iwo Jima, charging the enemy on foot.

Eye witnessed accounts of Tony crossing the mortar killing field without be killed or mortally wounded are recorded. Bare-chested, without a helmet or boots to slow his one-man barrage, his mind was made up to attack the fortifications of lurking Japanese forces single-handedly. The enemy would be shooting at him from an entrenched position, and he would be in plain sight, advancing regardlessly.

After he was well passed the killing zone at the beach head, he pressed towards the hillside. There was an interladen of formation that looked artificial, but danger could be prepared to strike from behind any crevice from which a rifle could be fired. From any lookout, the Japanese could view his approach, for they had been preparing for the Americans for some time. The cards were stacked against him for survival against such odds.

Corporal Tony Stein stood tall. Enemy fire echoed from a hill crest. The Japanese had taken the bait; they had revealed their stronghold position to him. Now he would charge the pillbox and defeat them. Victory, in this type of battle, would be reserved for the last man standing.

Tony charged forward to begin his offensive against obscured gunner and sniper attack. It was calm and controlled, unlike the Japanese kamikaze soldier in a desperate suicidal lunge at the enemy. The 1919 Browning guzzled ammunition, so he must fire the weapon with quick bursts, then change the belt without being shot as he reloaded. Since the automatic output of rounds were so rapid, Tony must constantly be aware of the amount of bullets fired, so not to run out at crucial times in the battle.

Corporal Stein squeezed the trigger that he personally improvised for a situation such as this. One, two, three, four, ... the report of rapid-fire blended to what sounded like one continual eruption. The enemy, in a feeble attempt at counter fire, volleyed back at him. The ground absorbed the bullets, as clumps of rock shot upward in his path. His forearms bulging to the weight of the weapon, he crouched to take better aim, and fired into the bunker port. The bullets found their mark. He changed the belt quickly, leaving the box of ammunition behind, slinging additional belts across his shoulders for his final annihilating charge at the Japanese defenders.

Tony raked the landscape and hillside with non-stop machinegun fire. He was close enough for the enemy to see his panthered arm glistening, they had never seen such markings on a soldier, only Japan's yakuza wore the tattoo so boldly. He must have appeared as surprisingly frightening to the enemy as could be imaginable, breaking the uniform code of which they strictly were to follow—without boots, helmet, and shirt—disobeying codes of uniform and personal conduct—carrying a portable machinegun with a 1200-1400 rounds per minute rate of fire. Worse yet, their bullets and mortars could not stop his advance. Perhaps they thought of him as one of the island's spirits taking human form. He was close enough for them to see the angry panthered forearm, cradling the weapon of their violent deaths.

Two Japanese attempted to attack from outside the bunker; they ran towards him from an exit hatch to the side of the fortified pillbox. One carried a sword, shouting as he attacked. Corporal Stein sprayed the rifleman alongside of him with a blast that incapacitated both; severing the hand off of the screaming swordsman. He alertly noticed movement from within the pillbox from the porthole, swung his weapon to the aperture opening, firing away into the shadowed figures inside. One of the enemy turned to shoot him with a handgun but shot his companion soldier, as his pistol jolted to the impact of the Browning 1919. The noise of automatic fire blended with the yelling Japanese to create the sound of chaotic warfare that is forever engrained in the mind, a clamor that for years will awaken from rest those to survive the Pacific War.

When the Shootout was over, the enemy lay silent. He must now return back across the killing fields for more ammunition. There was no end to this, only slight pauses in the continual bloodshed. Tony would cross this shooting range and mortar zone more than sixteen times, making at least eight trips across the concentrated machinegun fire and constant mortar barrage. His private D-Day invasion of Iwo Jima will never be forgotten.

Customizing His Weapon:

Detaching ANM Browning 1919 from aircraft
- Douglas Dauntless Torpedo Bomber
Navy Fighter Plane
Adapting it for land combat
Remove dual gips
Improvise trigger assembly
Replace with wooden sock of M1 rifle
Add rear sight and bipod from B.A.R rifle
Add front sight of M1 rifle
Reduce length of barrel for mobility

Corporal Tony Stein would rely on his skills
as a toolmaker to modify the machinegun
for a hand-held, personally adapted
weapon.

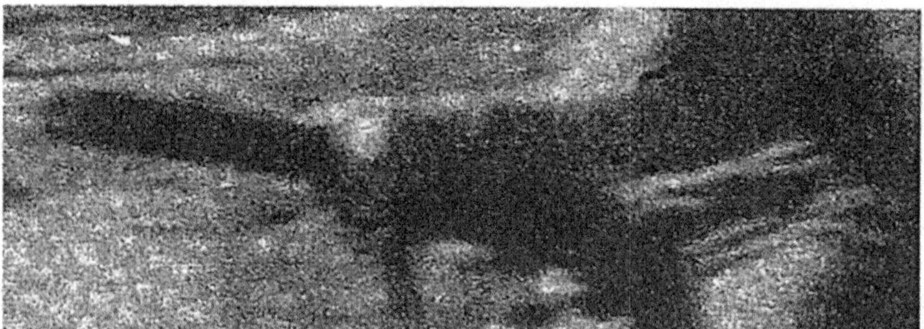

1919 Browning AN/M2 removed from a crashed U.S. Navy Plane blazing in the grasp of Corporal Tony Stein in combat at Iwo Jima, February 19th, 1945

After traveling back to the beachhead, Corporal Stein would reunite with a platoon of soldiers, and again set out to locate and destroy the hostile Japanese forces on the island. Since he was now somewhat familiar with identification of the pillbox fortifications, he would lead the way. His plan of attack was to flush out the enemy by causing them to shoot from their hidden whereabouts. The soldiers would slant back, allowing Corporal Stein to flank closer from one side of the suspected whereabouts of the bunker. When the Japanese opened fire, he would close in on them alone.

The enemy, however, knew how to wait. When the shots were fired, two gunners fired this time, one at them, and another from a cliff, directly at his back. He wheeled around, sprayed the cliffside with a burst of about twenty-five rounds, almost instantaneously. His platoon had also begun firing at the entrenched bunker. When the Japanese weapon nicknamed "The Woodpecker" began its clatter of automatic shooting, Corporal Stein revealed his position by standing in the open, as to draw attention away from his men. The Japanese then began to fire in his direction. He fired a quick burst and noticed movement from the side of his eye, from the same position of the cliffside assault. Corporal Stein turned and blasted the area with his adapted aircraft wing gun, kicking up smoke and rock fragments into the ravine, along with a falling Japanese sniper.

The Japanese bunker continued to fire his way while the platoon attempted to draw concentration back in their direction; but since Corporal Stein was closer to the enemy, he remained a greater threat. Tony had to reload his weapon and quickly changed ammunition belts while being under heavy machinegun attack. Almost immediately, he began to charge and return fire. Corporal Stein was again out in the open, testing fate, heading towards relentless automatic bullets from an entrenched bunker pillbox.

The Browning 1919 angrily growled and sparked led from its red-hot muzzle. Corporal Stein's arms tensed as they carried the thirty-one pound aircraft machinegun with its twenty-five pound hundred-round belt. To dash across a mortar field carrying such a monstrosity and advance towards a hidden fortification without being struck by enemy fire is nothing less than a miracle of Divine Intervention. Tony did this eight times.

As he drew closer to the Japanese hidden position, the bunker exploded in flames. The clay and makeshift roof of the pillbox had been blown away, revealing the burning Japanese. Two of the enemy were dismembered in place while operating their mounted machinegun; another three ran from the burning, entrenched oven. The platoon picked them off from a crossfire angle as they scattered from the pillbox, which glowed with the heat of an open furnace. The enemy fell to the ground, and then thrashed about in sporadic movements, to the sound of snap and sizzle. Such was the abomination of war.

The half-track was tank machinery that the men could use against stubborn pillbox strongholds. Some bunkers were built better than others, and rifles could not budge against them, since the enemy was entrenched in their confines. The situation would simply cost more American lives. The soldiers would have to approach these fortified ambush traps, to flush out the enemy without the help of an armored vehicle, and put themselves at high risk for mortality. So the armored vehicle was a welcomed sight in an island battle that at most times, could not afford to hesitate for one to arrive.

The half-track was moving against a complex of joining bunkers, when Corporal Stein directed their counter-offensive. Mortars were being lofted from the embankment of a pillbox which had a hilltop peak to provide concealment. When the platoon was pinned down, Corporal Stein would spray the bunker to divert the Japanese forces. The rapid fire of his weapon would cause them to hesitate, giving him the time to launch his personal attack, while the platoon could change positions. As the platoon and half-track worked on one bunker, he destroyed the other. His Browning 1919 cut the enemy to pieces at close range and when he could get to closer quarters with the fight raging on from now two other directions, his impact was even more devastating. The mortar launchers were hit with a barrage from his enraged weapon, one toppling forward onto the live prepared mortar, the body flying upward seven feet as it took the entire blast of the shell. The close combat was explicit and lighting fast, reflex seconds in action lasted an eternity.

The Japanese had been blasted down. Body parts and internal organs smeared to the honed-out hut extension of the tattered remains of bunker. Humans reduced to mangled flesh, but to pause for this type of consideration could cause one to lose one's life. It was not uncommon for the wounded enemy to lie in wait to set off a grenade. The Japanese were just as dangerous in an immobile state.

As the men of the platoon began to evacuate to their company position along with the halftrack, a distant enemy heavy machinegun opened fire. Corporal Stein had stayed behind, anticipating the Japanese to strike. He arose and began to fire; providing cover for the men of his platoon and acting as a rear guard for the armored vehicle. After his platoon was out of range, he gave the enemy one, last continual, savage burst of automatic fire. The Japanese heavy machinegun paused as if to pay homage to Tony Stein's personally adapted weapon. He had single-handedly dealt a private open attack which rivaled their sneak attack of Pearl Harbor. He had obliterated pillboxes and bunkers alike—a one-man demolition team of D-Day at Iwo Jima. The entire Japanese complex lay silent. There was only quiet to the oncoming evening of the dying and the dead.

The day was turning to night on the Pacific Island of nightmare and paradise. So much natural beauty of ocean and sky mixed with death and human suffering of war. The Marines had witnessed more that day than most do in a lifetime; and because of this, even the living survivors that returned back to the United States, would never truly leave the island. Their experience would cause them to forever linger on the confines, as if their spirits remained, sending their lifeless bodies home. Part of their entities seemed caught in between two worlds—the one they exist in, and the one they narrowly escaped from.

The red glowing horizon of the setting sun.

Tony looked to the West and at the glow of the omniscient sunset. Serenity had been briefly restored to this island of solemn tranquility. The Marines fell silent as the dead. The tropical breezes pushed the setting sun to its grave on the uncertain horizon. Twilight had given its final serenade.

Night had fallen on Iwo Jima.

On March 1st, less than two weeks after his heroic stand against the enemy, after volunteering to join a group of soldiers to locate a Japanese emplacement pinning down an entire Company, Corporal Tony Stein was met by a hail of machinegun bullets. Seven of the twenty that set out on the mission survived. Nearly seven thousand Marines were killed in-action on this island of Iwo Jima.

On December 16th, 1948, after being buried at the base of Mount Suribachi for three years, Tony Stein was returned to Dayton and rests at Calvary Cemetery.

His personal assault on D-Day February 19th, 1945 is perhaps the most courageous and awesome single-handed counter-attack by any soldier in combat.

February 19th, 1945
Afternoon of Invasion of Iwo Jima
First wave attack departing reporter
To arriving news correspondent:

I wouldn't go in there, (Iwo Jima) if I were you,
There is more hell in there
than in the rest of this War put together!

Note to Hollywood:

Your movie heroics are make believe.
You can have your fake macho stunts.
You can keep The Duke.
There's no need for plastic heroes
When genuine ones have lived and died,
Why invent stories of bogus substitutes?

-Anonymous

To the enemy at Iwo Jima:

One Corporal standing upright
In plain view charging in the open,
As if to say, Here I am

Is worth more than thousands of
your lurking soldiers hiding
In wait for sneak attack.

-Anonymous

To the island of Iwo Jima:

You forsaken island of death and suffering,
I would sink you into the sea,
For taking him away from Us,
If I could, I would blast you into the ocean,
and off of the map
right off of the globe.

-Anonymous

SECTION 3
23 YEARS OF SUNLIGHT

CORPORAL TONY STEIN WAS MET
WITH A HAIL OF GUNFIRE WHILE IN
ACTION. HE WAS 23 YEARS OLD.

TONY STEIN WAS BORN on September 30th, 1921 in North Dayton Ohio to Steve and Rose Stein. His parents emigrated to America in 1914; his father from Austria and his mother from Yugoslavia. He lived at 304 Alaska Street and attended Kiser High School. Stein, at age 19, saved the life of a drowning boy in the summer of 1940, documented in the Dayton area newspapers. This was the first of the many American lives that he would later save.

He worked to support his mom and family at occupations as the Civilian Conservation Corp during the Depression and Delco, where he honed his skills as a toolmaker he had started at a shop on Patterson Field. This ability would be applied to a crashed aircraft weapon in the South Pacific.

On September 22,1942, Tony enlisted in the United States Maine Corps against his mother's wishes. His nationalistic pride caused him to join along with 30,000 Montgomery County men and women who served in the armed forces during World War II. His mother wanted to keep him from the war but Tony Stein was compelled to fight for his country after the Japanese attack of Pearl Harbor. It was fitting that the boy with no fear of leaping into Dayton's Mad River to save the life of a drowning child would enter parachute training at Camp Gillespie, San Diego, when he completed basic training.

Breaking his leg on his first jump as a Marine paratrooper during training would not slow him down. Stein became a paramarine in the Marine Parachute Regiment. He would later become part of the 28th Marine division.

The Marine Corps train their soldiers to be more than adequate with a rifle, and Tony was an expert rifleman. In combat, on Bougainville during battle, he shot five snipers hidden in the trees. He was said to be, occasionally, in close quarter fighting with this type of dangerous enemy which would at times, strap themselves to the top of palm trees. Tony would sense any movement atop the trees, he told others, even the dashing eye movement of the enemy snipers. On Bougainville during the battle of Empress, Augusta Bay, he saved the life of his commanding officer. Corporal Tony Stein and the officer were near a group of trees, not expecting a sniper attack, when Tony suddenly threw the colonel to the ground and fired one shot into a nearby treetop. A Japanese sniper fell dead from the tree. Tony had either located the sniper by a keen perception or slight movement that gave the enemy whereabouts away. Corporal Tony Stein was to survive the heavy fighting at Bougainville, Vella Lavella, and the savage combat at Guadalcanal.

Tony Stein returned back to the north Dayton neighborhood that he loved in 1944 for R&R. He was married in that brief time to his wife Joan, an employee at Patterson Field, and went to San Diego for a three-day honeymoon. In June of 1944, Stein's sister Theresa commented about Tony's sense of commitment to return to the war. His wife Joan saw him for the last time in July of 1944, when he went to Hawaii for the special training that troops were to undergo before shipping to Iwo Jima.

The mission was for transport ships to carry his 5th Division Marines to the southernmost point of the island. Once there, under any attack from the Japanese from hidden locations that had been well-prepared, they must establish a beachhead for resupply. The Marines must then breakout from that point, to cut off Mount Suribachi from the rest of the island, and branch in forking directions. The objective of the landing at Iwo Jima was to secure the island from the Japanese, so it could serve as a landing base for aircraft and bombers, which could attack Japan from its important strategic location in the Pacific.

The type of fighting and amount of bloody carnage the soldiers were about to experience was unimaginable.

Corporal Tony Stein was first of his unit to be on station after hitting the shore of Iwo Jima. He was carrying his personally customized and modified weapon. During the landing at the beach, his Company A lost all commanders except for one. Fellow Marines were in awe at what Corporal Tony Stein was accomplishing under consistent mortar bombing and merciless Japanese fire on the island. Tony removed his helmet for better field of view and his boots for easier traction along Iwo's volcanic sands. Mortar shells exploded around him from all sides, as he traveled across the deep graveled-ash.

His valor to charge the Japanese pillboxes one by one, out in the open, is unmatched. To destroy an entire complex single-handed, while drawing sniper and machinegun fire in his direction to protect the others of his platoon, demonstrates the highest quality of loyalty. To run across an open field mortar attack sixteen times and only be struck non-critically on the seventeenth dash along such a killing zone is nothing less than incredible. To stand in the open against entrenched Japanese gunners and snipers in a shootout with a bulky aircraft weapon without being killed, and succeed in taking out eight bunkers, is much more than unbelievable, it seems an instance of Divine Intervention.

Once on Iwo Jima, it was the 28th Marines that must head towards the island stronghold—Mount Suribachi. Their dangerous detail was to capture the well-fortified volcanic mountain, which elevated to heights of 550 feet, at a near vertical incline, without cover of vegetation or any environmental landings to aid their ascension skyward.

Corporal Tony Stein kept a cool composure during the murderous machinegun fire and mortar bombing. In the face of terrific odds of mortal danger, he continued the covering of the platoon, drawing risk his way to alleviate the concentration of bullets in their direction. The ultimate destruction of the entire Japanese compound of pillboxes and bunkers must be attained before the soldiers of the 28th Marines could head towards Mount Suribachi. Corporal Tony Stein would see to this occurring under merciless attack of exploding mortars and relentless Japanese bullets, even though his weapon was twice shot from out of his grasp. He kept calm and continued sound judgment with a controlled fighting under intense fire and confusion of battle. Tony boldly charged the bunkers while bullets and mortars fell around him. His aggressive initiative allowed him to make undaunted decisions in the chaos, never wavering, providing help to wounded Marines in the process. On his ninth trip to the beachfront, he was struck in the left shoulder by shrapnel. The Captain ordered him to evacuate for treatment and Tony Stein refused. He continued his one-man onslaught on the attacking Japanese forces. At the end of the day, he had personally extinguished over twenty of the enemy.

After the complete destruction of the Japanese fortification complex on February 19th, 1945, the battered and bombed island of Iwo Jima fell to solemn quiet. There was only the stillness of the dead. In the days to follow, many more lives were to be claimed. The rocky beach angered by human intrusion and bombardment, indifferent to human suffering, providing no sentiment for the cruelty of mankind to itself and to the formerly undisrupted environment.

On March 1st, a group of volunteers were needed when the unit became pinned down by Japanese snipers firing from within Mount Suribachi's caves. Corporal Tony Stein volunteered and of the twenty men, only seven survived.

The American Legion Post 619 on Webster Street in Dayton has been named after Corporal Tony Stein. On December 19th, 1970, the destroyer escort USS Tony Stein was launched from Seattle, Washington. The Tony Stein Memorial Bridge at Keowee and Valley Streets stands to honor his name and heroic sacrifice. For his actions on D-Day at Iwo Jima, February 19th, 1945, he was posthumously awarded the highest honor of The United States of America, in one of the longest and most detailed citations written to acclaim a veteran of WWII—The Congressional Medal of Honor.

SECTION 4
A POETIC LULLABY TO THE HEROIC MEN OF IWO JIMA

It was in early December of '41,
Thunder came from the land of the rising sun;
On that Sunday laced with beauty so fine,
Tranquility turned to the end of days for those so young;

I will return to the east to repay
Thought he, never to right the wrong
Nevertheless, I will repay.

He would fly on land,
With feet to the ground,
Carrying a weapon of the air
He would stand upright to charge the enemy on foot.

A human aircraft was he
On D-Day Iwo Jima, 19th day of the 2nd month of' 45
Avenging Pearl, bare-chested to the setting sun,
Bare-foot to the blackened ash
Against an enemy that would dare to sneak attack
With groups of planes and bombs
To plunder those unexpecting ones of mine,
I will stand alone, thought he,
Unwavering under the merciless bombarding,
Shells and bullets will not stop me today.

The silver planes flew,
From the distant rising sun;
On the last month of that year,
To reign on the helpless ones.

Flying wings would take the lives of many,
With bombs and bullets to descend;
Metal and flames fall from above,
To the anguish of the red, florescent sky.

He would leave his homeland for country and friend,
Travel from west to the eastern shores,
To the distant land so far from home,
While crystal waves caped a powder-blue sea.

Farewell to the West and to Dayton,
So far away as the years of fragile youth,
They've come to this island, so few to survive,
The dirge death-dance of an island called Iwo Jima.

Stand Alone,
There's a deep blue sky overshadowed by
bellowing bursts of smoke and fire.

Just withstand that sky so blue,
Sky blue first stationed to land in the mourn,
In the morning of Iwo.

I'll detach that Browning from the aircraft,
thought he, the plane will be me.
As if strengthened by the downed bomber pilot,
He'd stand in front of the bunkers and dare.

I will charge them to their surprise,
Just to assault the assaulters,
A paratrooper to fly without wings.
He would dive bomb them in single-handed fashion.

Tony Stein stands upright, guarding his platoon,
He would draw hostile rifle rounds his way,
Now to charge the firing squad on foot alone,
Controlled was he, under the confusion of battle.

Pillbox to pillbox bunker,
While mortars and bullets fall near,
He unleashes the weapon to honor the fallen.
Only one country can send one to defy death so proud

The enemy had been told
American were not as strong as he,
Yet he walked towards many
While the snipers remained hidden
He would not yield to them
One man against many
Standing upright to charge on foot
At the island of Iwo Jima
On this day, no harm could take him
His weapon twice shot from his hands
The roaring Browning 1919
Today he could not be sustained
Cool and courageous was he
On the 19th day of the second month
In the year of liberation for the West
His panther was the dragon to the Japanese
Only yakuza wore such markings on their arms
As if the fright of his invincibility was not enough
To forecast doom to the land to the East afar
He approaches with a portable hand-held weapon
Protected by an invisible force-field
The bravest of the brave would avenge
The sins of their emperor at Pearl
Against the bellowing smoke of the open sky.

Red, White, And Blue,
Those are the colors; shades of other colors,
Shades of blood red, untarnished white,
Loyal shocking deep blue.

In a sea of fathom-blue reaching upward,
To powder-white clouds turning to red- haze
The open sky is angry,
The volcano goddess is blood-thirsty.

After the bombs of artillery,
The shells of mortar,
Send fire onto the ground
To vaporize all.

To destroy the pillboxes, the compounds,
The enclosed bunker fortifications
Cannot be leveled by bombs,
Planes will not take this island.

This soldier will find a weapon,
From a fallen bird-plane
Lift off a Browning ANM
He'll repay the American way.

He is going to stand tall,
Walk right up to them and show,
How an American fights,
Not like they did three years ago at Pearl.

Unyielding and non-stop,
Cool and courageous, Unrestrainable;
Indomitable and one-man against all,
This is the way.

Spilling led at hundreds of rounds per minute,
Empowered by America and Their Brave Men,
Spirits of the fallen Marines and Airmen,
His weapon from Their downed aircraft as tribute.

I will repay,
This Corporal has come from The United States,
Remember who you attacked in '41,
From the sea to the island of Iwo Jima.

He arrives to fight for the men of land, sea and air
Who fly to their deaths, up in the shattered sky,
Way up there in the clouds, in the powder-white clouds
In the clouds of crystallized dreams.

One man approaches the Japanese pillbox,
How can one stand in plain view and not be cut down?
A fortified closed bunker cannot defend them,
From Iwo's one-man assault.

One man walks toward them spraying led,
The enemy is strangely aware,
That bombs and bullets will not touch him.
He is protected by a force beyond.

They could not shoot him down today,
Perhaps they accept their fate when they see the panther,
On his arm as he sprays his led punishment,
Of belt-fed powered rounds per minute.

The island might take his life on some other day,
But today he is invincible,
He would fly on land with feet to the ground,
Look at what has arrived from the West,
Nothing can stop the indomitable Corporal Tony Stein.

While the rising sun awaits its doom,
The West will cradle the setting sun;
Your battle for the freedom fight,
Is soon over but never done.

One man advances towards the enemy,
Entrenched and waiting for him;
A closed bunker fortified pillbox
Firing rounds from concealment.

They cannot shoot him down
As he draws closer to them,
Spraying his led hailstorm
Of belt-fed onslaught.

Detached it from the fighter plane,
Modified the trigger, adapted for land battle,
The paratrooper and the aircraft are one,
Soldier and navy dive bomber are here to avenge.

He does not yield, protected by a force beyond,
The island may claim his life on some other day,
Wielding his AN/M2, more than twenty of the enemy
Would fall to the sound of his Browning 1919.

He is close enough for the enemy to see the panther,
Swift as if flying with the wings of the fallen aircraft
Guided by an intervention from beyond
The paratrooper, aircraft, and stinger all in one
Here is the human fighter-plane,
Soldier from the Shining Star.
Look and see what has arrived from the Western Sky.

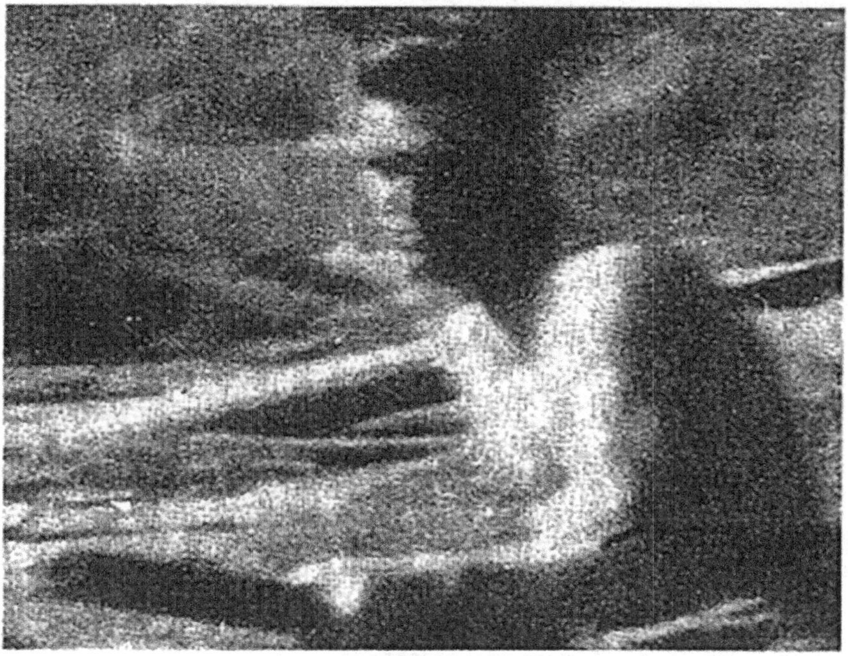

He cannot fly with the tiger,
So he takes the teeth from the Wildcat,
To run with the charging panther,
Until the bunkers fall silent to the snarling scream.

The nineteenth day of the second month is almost gone,
The wake of the day towards sunset,
He stands while his men watch on.
Looking up to the sky with his back to the east,
Beholding the sunset as the night falls.

Looking out to the western sky,
To the place where he had saved a child so long ago,
A life saved by him when he was aged nineteen,
Away to memories of the time,
    The splashing of the gentle tide
        onto the sands of Iwo
    The blackened ash of now blood-stained ground
        To the awestruck evening as darkness lends a hush.

Look out towards the setting sun,
You have fought beside the panther,
One man stands against so many and still lives,
His honor recognized to the reward of Intervention Divine.

Steps up this man against many,
With their mortars, pillboxes and bunkers to hide,
While he charges alone, standing in plain sight,
At the end of the day, he still will not fall.

It was the second month of the nineteenth day of his victory
For them taking the lives away in the sneak attack at Pearl,
He will fight against many whom crouch behind lies,
That now try to hide behind bunkers fortified,
Not shielded from his wrath in compartments concealed.

He stands in full view of his enemies,
Their bullets cannot touch him,
Nothing can bring harm to him on this 19th day,
The enemy fears one not afraid to stand alone.

He remains alive that day
No hostile fire can stop him,
While the unseen foe kills those who fall,
He remains alive when the sun rests to the setting sea.

Unharmed to the enemy,
He is shielded by a force Divine,
Given homage by the setting sun,
To honor him from his home country beloved.

February will not take you,
The second month is yours to live,
To fight at this distant place,
A near eight-mile rock of old.

He thought to himself:
I wish all this burning smoke
Wasn't here to blot out such a pale-blue sky;
Silence as dawn quickly fell to sunset.

The Pacific sun bowed to his silhouette,
His unmatched courage and skill in the chaos,
He was calm and unyielding while under merciless attack,
As he greeted the quiet of the looming nightfall.

The lavender surrounding highlight of the setting sun,
Did not hint of that day's mortar shells that fell,
Yet he was alive to the indifferent beauty of nature,
His profile outlined the beach for the pacific to awe.

As the 19th day of the second month came to end,
Sun sets on the island of doom,
Thousands would be claimed in time to follow,
Darkness will soon fall, yet live or die
The night will never have a claim to you;

Corporal Tony Stein yet waves no goodbye.

Swirling clouds and winds,
The passing years of change
Do not erase you from my memory.

You would have enjoyed the years to follow,
Taken away by an island of timelessness,
No civilization or life on that barren place,
No one born on that rock, there is only death in battle.

No green of vegetation only volcanic pebble,
Black, sandy quicksand
Does time take hold to the clash of combat,
Men bound to fight to the death …
What if countries begin to forget your sacrifice?

Where do you go to, my gallant friend?
The platoon you once saved on the nineteenth day
Glanced away for a moment and you were not there,
How strong you were in battle,
Standing firm and tall in your single-handed barrage,
With a blazing advance that would yield to none,
While bullets and mortar could not make you fall,
After that day, whatever will become of you?

The few that would survive have you in their memory still.
Standing upright with the aircraft Browning,
Belt of fire and sparking gold.

At the threshold of Suribachi
He waits,
As the months go to seasons past,
In August, a quake to shake the earth,
This island would shake like never before
More than it had five months ago,
Atomic was the sound from the east,
It signaled to the end of war;
Iwo was to gasp and still
On this island,
Those have fallen not so long ago.

Next comes the onset of autumn
In this far off land,
This land by the sea,
This rock of rocks,
Then the winds on winter gale,
To the next March of sorrow.

He would rest here a fortnight of years,
At the base of Suribachi,
Where did all the men of arms remain—
On here, this black ash of volcanic sand.
Why do they not clash, have they forgotten
The cause? Or are they in dismay?
Or do they merge with comradery for all—
Only the four winds know for sure.

Thousands have been taken on this island,
This island woe by the sea,
They salute to this one who will fight the many,
To lore that will have to forever last
As the pacific breeze whispers a final farewell.

He will, at last, return to home,
In the year of forty- eight,
To the land of his rebirth,
On the far and distant shore
Of yesterday's tomorrow,
Three Christmas Eves are long enough,
Until he is home again.

To sleep and dream of death,
Only to awaken by love to die;
In the flower of youth,
Do not call my courage to your door.

For once he saw sunlight
Be on that river lake,
And save the life of that child,
Back at the place of his birth.

Do not wait for his return,
That island rock will be to rest,
Spirits will soar from that land,
The body will be returned to home.

For on that day of winter nineteenth,
Only he would run with the panther.
To fight as one against all odds and prevail,
Nothing can harm one of Intervention Divine.

There are sounds that no one will hear,
In forgotten streams of silence,
Glimpses of visions long past,
While others fade, his stands alone.

Iwo Jima's one-man demolition personal assault on the enemy
The Unstoppable Corporal Tony Stein,
D-Day, February 19th, 1945

Nothing green, no life on this island rock.
No living things to roam this barren ground;
Land encircled by the pale-blue turquoise ocean,
So foreboding is this beauty of the Pacific Sea.

Many and many a year ago,
In a time most have long forgotten.
Thousands would come here to die in youth;
For ideals that sometimes seem to fade away.

Sweet year of youth,
Why have you come here to die?
Where no trees can or will exist,
On this island rock beyond the deep blue sea.

Where do these restless spirits go?
On nights of the howling wind.
In a land of silence before and since,
In the second month of woe by this turquoise sea.

Do not cry for the ones of valor,
Where on the island they remain
Unafraid to the setting sun;
Will you dream of them and me?

In the year of forty-five,
No angel to sound trumpets;
Of Armageddon to those,
In the land dead by the ocean sea.

Lest they never forget,
What the tides could never wash away,
To have been taken in this year of youth,
A cost with no witness to a heralded parade.

On the ground of blackened volcanic ash,
In the second month of forever,
Here flowers will never bloom,
That is, my friend, except for you.

And the whispers on the wind,
Still in rose-colored coral waves of sound;
Why have thousands come here long ago?
On this island rock by the crystal sea.

Gone is the sweet bird of youth,
To a place where no life will be;
Except for that one month of time,
For reasons lost to the whistling wind.

A sound from afar in August of that same year is heard,
On that island rock near the land of the rising sun;
To lift those waters of tomorrows turbulent sea,
And quake that island so indifferent to those fallen.

Who is it that will wait for you forevermore?
On that island called Earth by the baby-blue sea,
Goodbye for now; farewell to the setting sun,
In the land that since then birds will never sing.

Wait there by the shore of the crystal-blue ocean,
In the sweet year of youth you will be,
As we listen for the voices
Of those taken by the sea,
Just to hear the tides break,
By the beautiful sea.

Fading light dims the sight,
And a star gems the sky,
Gleaming bright…
From afar drawing nigh,
Falls the night.

Day is done, gone the sun,
From the lake, From the hills,
From the sky,
All is well, safely rest,
God is nigh.

Then good night, Peaceful night,
Till the light of the dawn,
Shineth bright,
God is near, do not fear,
Friend, good night.

-Anonymous

I will walk with you in the Spirit Realm

To be your guide
And protect you with the fury
Unmatched by any of this Earth
You are the Bravest of the brave
A Hero amongst heroes
Corporal Tony Stein
Your Courage lives Forever

And Always in my heart

In memory of
Corporal Tony Stein

Made in the USA
Monee, IL
04 November 2026

67374655R00056